Earth *to* Earth

violence of winds. They might also be placed as a shelter to the congregation assembling before the church doors were opened...' However, this use of yews for shelter does not really accord with their distribution because churchyard yews are significantly more common in the sheltered churchyards of the south of England and in Wales than in more exposed sites in the north, especially the north-east where they are particularly scarce. In Scotland the distribution has been claimed, for no very clear reason, to reflect the location of churches dedicated to the obscure fifth-century bishop St Ninian.

The observance of Palm Sunday, the Sunday before Easter when palm fronds were laid before Christ's entry into Jerusalem, has been traced back to the fifth or sixth century and it soon became the practice, especially in the Roman Catholic and Eastern Orthodox traditions, for palm branches to be blessed and carried in procession. However, as most real palms are too tender to thrive in northern climates, substitutes were required and willow branches with catkins were commonly used. When Easter falls early and the catkins are not yet out, yew has often provided an alternative. In some areas the day is even called Yew Sunday and in parts of Ireland the yew tree is often called a palm. At Wells cathedral in Somerset, a fine old yew stands in the centre of the cloisters, in what is known appropriately as the Palm Courtyard.

For reason that we have none Olyve that beareth
green leaf therefore we take Ewe instead of Palm
and Olyve and baren about in procession.

DIRECTION FOR KEEPING FEASTS ALL THE YEAR,
WILLIAM CAXTON, 1483

However, any evidence for yews being specifically
planted for this purpose is elusive. Moreover, any tree
around or over 2,000 years of age cannot of course have
been planted in a Christian churchyard, so the very oldest
trees such as that at Fortingall must have been growing
there before the church was built, and indeed before the
beginnings of Christianity itself. Did Christianity adopt
some earlier, pagan belief? It seems unlikely we shall ever
know and folk will continue to speculate, but also to admire
and stand in awe and reverence before these magnificent if
sometimes sinister plants, living calendars of history but
without any dates.

Chapter Five

LICHENS

As well in this wild kind as in planted Plum trees
of the hortyard, there is to be found a certain
skinny gum, in Greek called Lichen, which hath a
wonderful operation to cure the rhagadies, fissures
and chaps in the feet.

THE HISTORIE OF THE WORLD, COMMONLY CALLED
THE NATURALL HISTORIE II, P. 169,
PLINY THE ELDER, TRANS. PHILOMEN HOLLAND, 1601

Another kind of Lichen or Liverwort there is,
cleaving wholly fast upon rockes and stones in
manner of moss.

THE HISTORIE OF THE WORLD, COMMONLY CALLED
THE NATURALL HISTORIE II, P. 245
PLINY THE ELDER, TRANS. PHILOMEN HOLLAND, 1601

Thus, the name lichen was introduced by the translator
Philomen Holland to a largely unsuspecting English-
speaking world at the beginning of the seventeenth
century. Holland's references reveal the confusion that

has been attached down the ages to these remarkable organisms and their place in the wider scheme of things. It is a confusion that began with Pliny the Elder and is lamentably perpetuated by the Oxford English Dictionary to this day. For the Dictionary considers that the first quotation refers to lichens as modern science defines them, while the second refers to liverworts. I believe it is the other way round because although it is not at all clear to me what Pliny and Holland meant by the 'skinny gum' on plum trees, it is patently to lichens and not liverworts that the second definition applies.

It is entirely understandable that lichens (far better pronounced 'liken' than 'litchen') were confused with liverworts because the superficial appearance of some forms of both can be confusing to the untutored eye; and both in turn can be confused with some kinds of algae, some kinds of moss, some kinds of flowering plant and some kinds of fungi. In Britain, far behind Pliny in its understanding of the natural world, the earliest indisputable reference to lichens came at the beginning of the tenth century when Anglo-Saxon documents described what were obviously lichen-covered trees using the word 'har', the origin of the modern word 'hoar', as in 'hoar-frost'. However, for many centuries, apart from casual observations like these, almost all interest in plants and plant-like growths was the preserve of herbalists who used them to greater or lesser effect to create their medicinal remedies.

Things started to change, however, from the sixteenth

century onwards, when an interest in plants for their own sake fostered the expansion of the science of botany. Botanists began to collect lichens, dry them and include them in their herbarium collections, to give them scientific names and illustrate them in their books and journals, although still without knowing what they were and how they related to other living things.

For centuries it has been obvious that churchyards, and especially the gravestones they contain, offer as rich a lichen hunting ground as anywhere; that lichens and gravestones form an almost obligatory partnership:

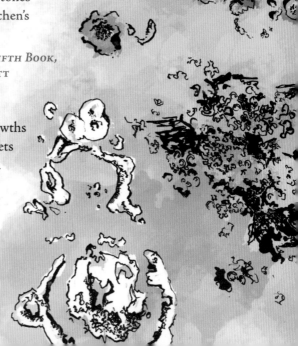

> ... many a fervid man
> Writes books as cold and
> flat as grave-yard stones
> From which the lichen's
> scraped...
>
> AURORA LEIGH – FIFTH BOOK,
> ELIZABETH BARRETT
> BROWNING, 1856

What are these growths that have inspired poets but for so long baffled herbalists, naturalists and casual observers alike? It is of little

surprise that folk were puzzled because, uniquely in the living world, lichens are two quite separate organisms growing together and acting as one, a state of affairs that could only properly be elucidated once microscopes became sufficiently sophisticated in the mid-nineteenth century. So the reality was at last revealed: lichens comprise a fungus and an alga or a green alga-like growth in close, obligatory harmony. Sometimes it seems to be of mutual benefit; in others the fungus is parasitic on the alga.

Aside from their fascinating and extraordinary lifestyle that so appeals to biologists, lichens can be admired and appreciated by everyone, as they exist in astonishing variety of size, shape, form and colour, and are often exquisitely beautiful. Experts in lichens – lichenologists – recognise six or seven different forms of lichen growth but for the non-expert it is easiest to appreciate three main types, each with a descriptive name and all of which can be found in churchyards: crusty lichens, leafy lichens and shrubby lichens. They have the distinct advantage as subjects for admiration and study that unlike most plants and fungi they appear the same all year round and, in fact, often stand out particularly beautifully when illuminated in an otherwise bleak landscape by a low winter sun.

Beauty, beauty? What is it? Is it only a trick
of old stone and lichen in sunlight?

THE DIARY OF ARTHUR, CHRISTOPHER BENSON, 1926

Why churchyards? First, because lichens are slow growing, in some cases extremely slow growing – perhaps 1 mm per year – and so more than almost every other living thing in this book, they benefit from the lack of disturbance offered by a churchyard. The stones of the church itself and the boundary wall commonly remain untouched for centuries; trees in churchyards too may be seldom subject to routine pruning or replacement. Most importantly however, within that stable and protected environment, it is the gravestones and chest tombs with their horizontal moisture-retaining covers that above all offer the perfect platform for lichens to colonise. Lichens are the most important colonisers of virgin rock anywhere because they have the ability both to adhere to the surface and to obtain nutrient from the minerals within. Thus, they can grow without the need for soil and in the wider natural environment are hugely important in actually beginning soil formation from newly exposed rocks.

The rougher the rock surface, the easier it is for lichens to gain an initial foothold, so they are most common and abundant on older gravestones, especially those made from sandstones and grits which also contain readily accessible minerals. On gravestones made from harder rocks such as granite or highly polished ones such as marble, lichens will be fewer and less diverse. Moreover, gravestones are often made from types of stone different from those occurring naturally in the area; limestone gravestones, for instance, may be found where the local rock is granite, so a graveyard

may be in effect an oasis habitat. Of course throughout much of lowland Britain, because there are no natural outcrops of rock of any kind, graveyards then become even more significant as lichen habitats and may be the only places for many miles where certain species occur. Of the 1,700 or so British lichen species, over 300 have been found growing on churchyard stone in lowland England and many churchyards contain well over one hundred species. Almost half the churchyard species are rare, having been found in fewer than ten sites, while some seldom occur in any other habitat.

Churchyards also contain fascinating micro-habitats. Close to church lightning conductors, for instance, rare copper-tolerant lichens may be found. Interestingly, and certainly until the passing of clean air legislation, country churchyards have supported richer lichen colonies than those in towns for the simple reason that many lichens are extremely sensitive to atmospheric pollution. In fact, so sensitive are they that the presence or absence of certain species can give a close indication of the amount of air pollution in the locality.

> Tactile memories
> in the cemetery stones
> visual reflections of the times
> the ages in which the country knew
> imagery of hell, of heaven,
> of lives as they were lived,

94

or of how they want to be remembered
tactile granite, marble, slate,
yielding in their turn to the rain,
the ice, the lichen, the moss,
the wind, the ages
Hear the whispers in the wind
the feel of the cemetery stones
rough, cool, smooth
places of remembrance
to be felt.

CEMETERY STONES, RAYMOND A. FOSS, 2007

Because their true nature has only been appreciated in recent times, many of the references in older literature to moss referred to lichen and not to real mosses, liverworts or algae. In *Cymbeline*, Shakespeare tells of Arviragus, who mourns the supposedly dead Imogen then masquerading as the boy Fidele, and promises to her that he will 'sweeten thy sad grave' with primroses, harebells and eglantine...

... Yea, and furr'd moss besides,
when flowers are none.

CYMBELINE, IV. 2

The 'furr'd moss' was surely the lichens that would grow on her gravestone. And I would not bet against lichen also being the subject when...

> Under an oak, whose boughs were moss'd with age
> And high bald top with dry antiquity,
> A wretched ragged man, o'ergrown with hair,
> Lay sleeping on his back...
>
> *As You Like It*, IV. 3

Shakespeare was less complimentary elsewhere:

Not marble, nor the gilded monuments
Of princes, shall outlive this powerful rhyme;
But you shall shine more bright in these contents
Than unswept stone, besmeer'd with sluttish slime.

SONNETS LV. SIG. D4

The shrubby kinds of lichen that grow on upland or moorland habitats may be grazed by sheep and deer while in Arctic regions they form a hugely important food resource – sometimes the only food resource – for reindeer and other native herbivorous mammals; hence the colloquial name 'reindeer moss' or 'Iceland moss' for one important species. Some insects and other invertebrates also feed on lichens while several birds, especially long-tailed tits, use them in nest construction. In Britain, lichens have seldom been a significant food source for humans but they have been used since ancient times as a hugely important source of natural dyes, most notably the red-brown or yellow-brown crottle dyes and the purple or red staining dyes orchil or cudbear. They once furnished a significant and profitable industry:

Crodh air na beanntan òr air clachan
[Cattle on the hills, Gold on the stones]

TRADITIONAL SCOTTISH SAYING,
ATTR. IAIN MACCODRUM 1693–1779

For over five centuries lichens have found medicinal applications of varying effectiveness and more recently it has been discovered that some contain substances such as usnic acid, which possess slight but valuable antibiotic properties. Some species have also found modest use as sources of anti-inflammatories, essential oils, preservatives, toxins, deodorants and perfumes. However, surely the most celebrated lichen product was the substance lichenin which prevented ageing to a quite remarkable degree. Sadly – or perhaps not – it was the product not just of a lichen but also of the remarkably fertile imagination of the distinguished science fiction author John Wyndham:

> Zephanie caught her breath suddenly, and turned round, almost fiercely. The tension in her whole body hardened in her voice. 'I asked you how long', she said. 'How long am I going to live?' Francis looked back... With a curious kind of pedantry flattening any emotion out of his voice, he replied 'I estimate your expectation of life, my dear, at approximately two hundred and twenty years.'
>
> TROUBLE WITH LICHEN, JOHN WYNDHAM, 1960

Like all other living things, lichens are named using the international binomial system based on Latin grammar and developed by the Swedish naturalist Carl Linné [Linnaeus] in the eighteenth century. Whilst invaluable in avoiding

ambiguity, these names can be cumbersome and awkward for the general public – *Sagiolechia rhexoblephara* and *Allantoparmelia alpicola*, for instance, while linguistically elegant, do not readily trip off the tongue. Thus, there has been a move in Britain in recent times to give all living things an agreed English name to sit alongside its scientific designation. Moreover, the Wildlife and Countryside Act requires all protected organisms to have an agreed English name so some have had to be invented. For whatever reason, this seems to have met with more opposition from lichenologists than most other biologists and there is still no universally agreed, approved list of English lichen names. As a result, while names such as the Southern Grey Physcia and Forked Hair-lichen have been invented for a few rare and protected species, even some really common types such as the extremely familar orange crusty lichen *Xanthoria parietina* do not have an agreed name and different authors have none too imaginatively called it Yellow wall lichen, Common Yellow Wall-moss, Common Yellow Wall-lichen, Maritime starburst lichen or Yellow crotal.

Lichens are valuable in the churchyard, not just in painting stone work with a palette of remarkably subtle beauty but because different species colonise different micro-ecological niches, so they can actually create contrast and enhance details. The incised wording on old gravestones is often rendered more visible because the lichen species growing in the slightly more shaded and damper letter incisions are different from those on the more exposed stone surface. Moreover, while the way churchyards are managed has a significant impact on the ability of most living things to thrive, what is good for some species may actually be bad for lichens. Leaving plant growth to flourish unchecked in order to create an appealing wildflower meadow environment, for instance, can result in gravestones becoming shaded and lichen growth suffering in consequence.

THE LICHEN ON THE WALLS

Ah! the lichen on the walls
Out in patches, white and red,
Where the creeping ivy crawls
O'er the tree-stems overhead,
Is a token that no throng
Bustles by them all day long,
Ever wearing roadways bare
As they drive the whirling air
Where the way
Is not footless, night or day.

In the ever-busy street,
Where we see no growing grass,
Streaming folk that briskly meet
Throng each other as they pass;
Rubbing wall-sides to a gloss
Where is little soil for moss
That can seldom linger quick
On the ever-fretted brick,
And no seed
Ever quickens to a weed.

Here the words of fewer men
Come with sounds of bird and bee,
And the mossy-nested wren
Flits from ivy on the tree.
And the open sunshine glows
On the lily or the rose,
Leaving shaded air to cool
In the bower, and on the pool·
And a sound
Under others is not drown'd.

There are sire and wife, now old,
And two sons, both hale and lithe;
And two maids of comely mould,
Who are ever kind and blithe.
And whenever maid and son
Shall have mated, one by one,
Then, as peaceful be the life
Of the husband and the wife,
As they all
Have within the garden wall.

WILLIAM BARNES, N.D.

Chapter Six

MAMMALS, REPTILES
AND AMPHIBIANS

It is possible, if not probable, that every species of wild British vertebrate – except presumably fish – has been found at some time in a churchyard. It is not my intention here to roll out a pointless list of them. Rather, I shall draw attention to those mammals and birds – and to a minor extent reptiles and amphibians – for which churchyards are especially significant habitats.

First, moles, which must be the most apparent churchyard mammals because the soft, undisturbed, earthworm-rich soil between the gravestones is so commonly pock-marked with the small mounds of ejected soil called mole-hills that betray their subterranean activity.

> Diggory Diggory Delvet!
> A little old man in black velvet;
> He digs and he delves –
> You can see for yourselves
> The mounds dug by Diggory Delvet.
>
> APPLEY DAPPLY'S NURSERY RHYMES,
> BEATRIX POTTER, 1917

Surely no animal is Britain is so well known but so infrequently seen, no creature so fascinating yet so misunderstood. Moles are despised and loathed by churchyard gardeners but adored by generations of children to whom the likes of Mole in Kenneth Grahame's *Wind in the Willows* and Moldy Warp in Alison Uttley's *Grey Rabbit* stories have long been an essential part of growing up. It is slightly anomalous therefore that in reality moles are vicious little creatures with poor eyesight and hearing but have fore-legs like excavators and ferocious teeth.

If moles are the most common churchyard mammals, rabbits are undeniably the most troublesome, grazing both the vegetation and the flowers placed on graves. There are in truth few plants that a hungry rabbit will not eat and while a traditional churchyard boundary wall offers an effective barrier to them, hedges are an open invitation and of course churchyard gates can hardly be made rabbit-proof. Left to their own devices, especially in closed and relatively undisturbed churchyards, they will burrow and may even disturb the graves.

Yet it is impossible to imagine Britain without rabbits. No other introduced mammal has so swiftly become completely integrated into our life and literature.

The primroses were over. Towards the edge of
the wood, where the ground became open and
sloped down to an old fence and a brambly ditch
beyond, only a few fading patches of pale yellow
still showed among the dog's mercury and oak-tree
roots. On the other side of the fence, the upper
part of the field was full of rabbit-holes.

WATERSHIP DOWN, RICHARD ADAMS, 1972

Yet rabbits are aliens. There are no archaeological remains
of them in Britain before the twelfth century, there is no
Anglo-Saxon name for them and they were not mentioned in
the *Domesday Book*. The conclusion is that they were brought
here for food and fur by the church-building Normans from
southern France and the Iberian Peninsula.

By the early 1950s the rabbit population of Britain was
vast: perhaps 100 million animals. It held an apparently
impregnable position as the most numerous of all British
mammals. However, rabbits were to suffer the most
catastrophic decline of any warm-blooded creature in these
islands in recorded history. A South American virus was
responsible and myxomatosis reached England in 1953.
Legislation was passed to prohibit the deliberate spread
of the disease and of affected animals, but to no avail. It
spread across the country like a wildfire and killed around
ninety-nine per cent of the rabbit population, although they

have now recovered significantly. Although rabbits form an important food source for several predators, including foxes, domestic cats and the larger birds of prey, their extremely effective reproduction, early sexual maturity and year-round breeding mean their numbers are now seldom seriously depleted.

The presence of badgers in a churchyard creates a unique problem because they are the only creatures that can cause significant physical damage but nonetheless enjoy legal protection. They live in social groups in often large and complex underground chambers called 'setts', accompanied by major above-ground earthworks. I have never seen a sett in a churchyard and they are usually sited in or close to deciduous woodland. Nonetheless, badgers will commonly enter churchyards where they can cause more limited but nonetheless significant damage and disturbance as they dig for food. Despite their prodigiously strong jaws and teeth, badgers are omnivores and will eat almost anything that comes their way, including carrion.

Apart from keeping boundary walls in good repair and closing churchyard gates securely at night – for badgers are largely nocturnal – there is little that can be done to dissuade them because badgers are shielded from harm and their setts from interference

by the Protection of Badgers Act 1992. Intended originally
to outlaw the hideous pastime of badger baiting, many now
see it as anachronistic, not least because in designated areas
badgers are lawfully killed in support of the belief that
they are responsible for the transmission of tuberculosis
in cattle.

> He hath very sharp teeth, and is therefore
> accounted a deep-biting beast: his legs (as some
> say) are longer on the right side than on the left,
> and therefore he runneth best when he getteth
> to the side of an hill.
>
> THE HISTORIE OF FOUR-FOOTED BEASTS,
> EDWARD TOPSELL, 1608

Collectively, one of the major subdivisions of rodents –
mice, rats and voles – are easily the most numerous British
mammals. The most common of them all, the field vole, occurs
in its millions almost everywhere and I doubt there is any
churchyard without a resident population. Apart from brown
rats which carry infectious diseases and are certainly not to
be encouraged, all the small rodents should be welcomed
and tolerated in churchyards where they are important
members of the soil level community, feeding omnivorously
on a wide range of organic matter and themselves providing
significant food for predators such as owls, weasels and
stoats; all in their turn common churchyard associates.

Wee, sleekit, cowran, tim'rous beastie,
O, what a panic's in thy breastie!
Thou need na start awa sae hasty,
    Wi' bickering brattle!
I wad be laith to rin an' chase thee,
    Wi' murdering pattle!

To a mouse, On turning her up in her Nest with the Plough, Robert Burns, November, 1785

Although they are quite unrelated to rodents, shrews are conveniently grouped with them as almost obligatory inhabitants of grassy places, including churchyards, and their numbers are also regulated by a similar range of predators for which they too provide food. Nonetheless, while domestic cats will kill them, they are often reluctant to eat shrews, presumably finding the liquid produced by certain skin glands to be distasteful.

In passing I might add that domestic cats should be added to the list of the most common churchyard mammals, where they do perform an invaluable role in regulating the rodent population. There is also a fascinating churchyard story relating to native wildcats which are now to be found only north of the Great Glen but in centuries past appeared throughout much of the British mainland. In or around 1477 a young man named Sir Percival Cresacre was returning from Doncaster through Melton Wood in south Yorkshire when he was allegedly attacked by a wildcat and sought refuge in the porch of St Peter's Church at

Barnburgh, where the fight continued. In the morning man and cat were found, the man dead of severe lacerations, the cat crushed against the wall by his feet.

However the mammals most frequently associated with churches and churchyards, certainly in popular perception, are bats, perhaps through the well-known expression 'bats in the belfry'. Apart from birds, they are the only warm-blooded creatures to have successfully and completely taken to the air. They have achieved this by an enormous extension of the finger bones to support thin flight membranes.

> The Reremouse or Bat, alone of all creatures that flie, bringeth forth young alive: and none but she of that kind hath wings made of pannicles or thin skins. She is the onely bird that suckleth her little ones with her paps, and giveth them milke; and those she will carrie about her two at once, embracing them as she flieth...
> No flying foule hath teeth, save only the bat or winged mousee.
>
> *The Historie of the World*, commonly called *The Naturall Historie II*, Pliny The Elder, Trans. Philomen Holland, 1601

When vesper trails her gown of grey
Across the lawn at six or seven
The diligent observer may
(Or may not) see, athwart the heaven,
An aimless rodent on the wing. Well that
Is (probably) a Bat.

THE DAY'S PLAY, A. A. MILNE, 1910

As distinguished a writer as he was, A. A. Milne was no zoologist. Bats are not rodents and when they fly, they are certainly not aimless. They use sophisticated echolocation to detect the flying insects on which they feed. Although they do have eyes – and are not 'as blind as a bat' – these seem to have limited use; not least because all our species are more or less nocturnal, a feature that has not helped their public appeal. Because, let us be honest, regrettably most people do not like bats and some are terrified of them.

Together with dormice and hedgehogs (another welcome churchyard mammal, now happily regaining some of its lost numbers) bats are the only British mammals to hibernate, although they will emerge and fly during warmer winter weather. Their winter hibernating places, or roosts, are often in some elusive underground site such as a cave or mine shaft, although our most common native bat, the pipistrelle, prefers buildings and vast numbers are sometimes found in barns and churches. The summer roosts of most bats by contrast are much more easily found and differ rather characteristically from species to species. Some species roost in trees but many do so in undisturbed places in buildings, sometimes in huge numbers and their unexpected discovery can come as a serious surprise. Their noisy squeaking may be heard many metres away and their droppings mount up in significant heaps.

The association of bats with churches is an old and well known one. Bats are unlikely to distinguish between a church and a barn as a suitable summer roost but, nonetheless, the more specific association of bats with belfries is as mysterious as it is misleading. Belfries are almost the last places in a church that you would expect to find bats, as they are just too cold and draughty. In the roof of the south porch is a much more likely spot. It is nonetheless surprising that someone as familiar with churches as John Betjeman should have been taken in; he called his 1945 anthology 'New Bats In Old Belfries'. I recently asked a bat expert if there was a correlation between bat populations in churches and the tenor of the liturgy: does the high church service with its incense deter them? There is apparently some evidence that it might.

Despite the extremely common occurrence of bats flying

over churchyards and being associated with religious places generally, there is also a long-standing and contrasting connection between bats and witches. In past times, it was believed that bats were witches in disguise, and in the famous scene in Macbeth, 'wool of bat' was an ingredient of the three witches' cauldron brew (Macbeth, iv. 1). The familiar expression 'like a bat out of hell' also draws on the unfortunate association between bats and evil.

There are around fourteen British bat species, however many are rare and most British bats are pipistrelles. All enjoy greater legal protection than any other mammal and neither the animals themselves nor their roosts may be disturbed or even touched by anyone without a special permit. The law now requires private individuals and ecclesiastical authorities to take account of this when planning renovation or new buildings. For those species that will adapt to them, the provision of bat boxes has become almost as familiar an activity as the erection of bird boxes in gardens and I have seen them in the more conservation consciously managed churchyards.

Finally, a note about reptiles and amphibians which are fairly uncommon in churchyards: adders and most lizards because they are shy of man and of interference, andgrass snakes, newts and frogs because of the absence of ponds. However the legless slow-worm seems to appreciate the peaceful environment and is rather common in churchyard compost bins, while toads quite frequently turn up in cool damp corners inside churches.

I shall soon be laid in the quiet grave –
thank God for the quiet grave – O! I can feel
the cold earth upon me – the daisies growing
over me – O for this quiet – it will be my first.

ATTR. JOHN KEATS, 1821

Chapter Seven

BIRDS AND SMALL CREATURES

Just as most types of British mammal have probably been found at some time in a churchyard, so I suspect the majority of our bird species must at some time have flown over one. My chief interest here, however, is in those birds that find the churchyard a particularly favourable habitat in which to feed and especially to breed.

As I mention in several contexts throughout the book, the churchyard habitat encapsulates either a bit of the surrounding environment or, rather commonly and more importantly, a bit of the environment that was present in times past. If the area of the churchyard is large enough – at least half a hectare or so – the plant and to some extent the small mammal population within it may be self-contained and self-sustaining. Significantly different criteria apply to birds, however, and other than with the largest churchyards and smallest birds, there is unlikely to be a bird population that stays put and never leaves. There are several reasons for this. Birds are highly mobile, many are nervous and intolerant of human interference or presence and, most significantly, most would not find sufficient food. So the birds of a churchyard will generally be the birds that have flown in temporarily from the surrounding area – farmland birds in a rural churchyard, woodland birds in a woodland churchyard, garden birds in a suburban one, pigeons in an inner city one and so forth.

Given these constraints, what will induce any individual bird species to linger, feed and nest? First, the presence of a food supply. For many species this will be linked with the all important churchyard plant life which may provide food directly in the form of foliage or seeds and fruits; and indirectly by encouraging an insect population. While many bird species, especially many small song birds for instance, are not choosey and feed on pretty well all and any available seeds or insects, some have more specific tastes. There are numerous examples: blackbirds eat many things but are especially fond of earthworms, so may be expected on and around closely mown areas of churchyard grass or mulched plant beds, whereas their close relatives thrushes, which have a particular predilection for snails, will often prefer dry-stone walls or long grass where snails are more likely to occur. Wrens eat many types of small insects and spiders but seem particularly attracted to plants infested with aphids, which they pick over meticulously.

Some birds have even more specific preferences. For example, many garden owners will be familiar with the fondness of goldfinches for thistle and similar seed heads, and of bullfinches for fruit buds. Yet other birds find churchyards a good food source for different reasons. Our most common owl species, the tawny owl and its less frequent relative the barn owl – sometimes called the church owl in the north of England – are good examples. Both may be seen hunting over churchyards at dusk, searching for voles, their preferred prey which occur in large numbers among the unmown grass. Sparrowhawks, one of the most common British birds of prey are comparably attracted by the presence of sparrows and other small song birds.

The second churchyard feature of importance in attracting birds is the suitability of nesting sites. It is not always appreciated that most bird species have fairly specific requirements for the positioning of their nests. Some, like many game birds, nest in what is little more than a scrape on the ground. Others, like skylarks, are also ground nesting but build a slightly more sophisticated nest. However, most songbirds – blackbirds, finches and warblers, for instance – nest in trees, shrubs or among herbaceous plants and may even exhibit a preference for specific plant species. Large birds like crows and rooks tend to nest high in trees, swallows in barns, martins on the outside of buildings under the eaves, owls and woodpeckers usually in holes in trees, and so on.

> The summer nests uncovered by autumn wind,
> Some torn, others dislodged, all dark,
> Everyone sees them: low or high in tree,
> Or hedge, or single bush, they hang like a mark.
>
> BIRDS' NESTS, EDWARD THOMAS, N.D.

Therefore, if the appropriate vegetation is present and the birds occur locally, a churchyard may provide suitable nesting sites for them, although a significant number of bird species are likely to be excluded because most churchyards are fairly sparsely furnished with trees and other than in some closed churchyards, ground nesting birds will be subjected to too much disturbance.

However, there are some less obvious options in the fabric of the church itself. The undersides of the numerous

ledges on more ornate churches may be used by martins to build their mud and plant material nests. The ledges themselves may be used by pigeons or sparrows simply because it is often impractical for each and every ledge to be furnished by the church authorities with deterrent devices. The interior of the church is an attractive nesting proposition for swallows and for opportunist and less fussy species like sparrows and pigeons, although the authorities generally take measures to prevent them from entering and soiling the fabric with their droppings – many churches have wire netting doors inside the porch accompanied by a request for visitors to keep all doors closed.

Historically, the most remarkable example of a church nesting site must be the pair of white storks – once an occasional British bird – that nested in 1416 on St Giles' Cathedral in Edinburgh. However, the most unexpected and most appreciated use of church – or more usually cathedral – buildings in recent years has been by birds of prey, especially peregrines. Where once a typical peregrine nesting site was a sea cliff, it is now almost as likely to be high on a cathedral ledge and many of our great cathedrals have installed 'nest-cam' monitoring of the sites with the birds' activity readily observable on-line.

> Hie you to church; I must another way,
> To fetch a ladder, by the which your love
> Must climb a bird's nest soon when it is dark…
>
> ROMEO AND JULIET, II. 5

Today, practically all birds and many mammals, reptiles and amphibians enjoy legal protection. It is a sad but understandable truth, however, that the smaller and less obvious or attractive a rare or threatened creature, the less likely it is to have such benefit. So apart from butterflies and dragonflies, rather few insects and other invertebrates, no matter how rare, are comparably protected. This means that by far the majority of churchyard inhabitants, from aphids to spiders, woodlice to centipedes, ants to eelworms pass unseen, unremarked and unknown, and it is generally more by good fortune than by being accorded special attention that they survive. Only in recent times, when the importance of churchyards as wildlife habitats has become more widely recognised and they have been accorded serious study, has it been possible to assess the significance of any of their invertebrate populations. I shall therefore simply draw attention to a few examples of those animals for which churchyards are – or may be assumed to be – significant habitats.

First, the woodlouse, the creature with more colloquial names than any other; over 200. One of the names is 'church pig', testimony to one of its favoured habitats! There are around 40 British species, many feed on rotting

wood and other organic matter, and the more common kinds have increased significantly in recent years. All share a requirement for cool, damp conditions. Lift a stone or old log and woodlice will be seen beneath it. It is this necessity that caused so many of them to wander from churchyards into churches; and to some degree it still does, although the advent of central heating has made many churches far less appealing for them than they once were.

Insects are a prodigiously large group of highly successful creatures. Among them is the most successful group of all: beetles. They comprise one fifth of the 20,000 or so British insect species. Undoubtedly, many occur from time to time in churchyards, although like many kinds of insect relatively few are seen by the casual visitor because so many are small and live well concealed lives among the vegetation at soil level. Among them is the only creature – as far as I know – that is actually named after the locality: *Rhizophagus parallelocollis*, the graveyard beetle. Although no doubt important in recycling, it is an altogether unwholesome beast because it is said to be 'common in cemeteries where it swarms in graves, around tombstones and in one to two year-old corpses'.

One group of insects, the butterflies, are of relatively minor importance in the wider scheme of biodiversity, albeit their caterpillar larvae

are voracious
feeders, but enjoy
a disproportionately high
public profile because of their
undoubted beauty. I suspect a fair
amount of 'wild flower' planting in
churchyards is as much to attract butterflies as it
is for the flowers themselves.

> I've watched you now a full half-hour,
> Self-poised upon that yellow flower...'

To a Butterfly,
William Wordsworth, 1801

There are around seventy British butterfly species, many declining in numbers and now rare. However, if the objective is to encourage butterflies to breed and proliferate, using flowers to attract them is only part of the story. Much more important is the presence of food plants for their caterpillar larvae.

If planting flowers to attract butterflies may be a little limited in its ambition, planting them to attract bees is not. Bees of most kinds are hugely important pollinators; horticulture and agriculture could not exist without them. Most important of all are honey bees but our twenty or so magnificent species of bumble bee are especially in need of encouragement and conservation, as so many have declined in recent years. The undisturbed

flora of churchyards is an ideal environment in which this can occur.

> There's a whisper down the field
> where the year has shot her yield,
> And the ricks stand grey to the sun,
> Singing: – 'Over then, come over,
> for the bee has quit the clover,
> And your English summer's done.'
>
> THE LONG TRAIL, RUDYARD KIPLING, 1891

Spiders have enjoyed an ancient association with churchyards. It has long been believed that for a bride and groom to see a spider on their way to church portends good luck. Among the 600 or so British species, there is a huge range in size up to the uncommon cardinal spider which can exceed 13 cm across the legs and sometimes appears in old buildings. It gained its name from giving apoplexy to no less a churchman than Cardinal Wolsey when he and it shared a residence at Hampton Court Palace. There is another curious connection of spiders with churchyards in a belief that certain types of wood were repellent to spiders and that they and their cobwebs would never be found in buildings made from them. At Winchester Cathedral, it was believed that spiders would not make webs on the chapel or cloister roofs because they were made from Irish oak, an old belief that has some connection with St Patrick

and his banishing of vermin from Ireland. Chestnut wood is said to be particularly distasteful to them and the chapel roofs at New College and Christ Church in Oxford are said to be always free from cobwebs.

My fairy lord, this must be done with haste,
For night's swift dragons cut the clouds full fast,
And yonder shines Aurora's harbinger;
At whose approach ghosts, wand'ring here and there,
Troop home to churchyards; damned spirit all,
That in crossways and floods have burial,
Already to their wormy beds are gone.

A Midsummer Night's Dream III. 3

Chapter Eight

THE FUTURE OF CHURCHYARDS

Horticulture's relationship with wildlife has changed in many ways over the past forty or fifty years and I believe it largely began with a book called *Silent Spring*, published by the American author Rachel Carson in 1962. She was particularly concerned about the widespread and almost unregulated commercial use of chemical pesticides that persisted for many years in the environment with serious consequences for wild creatures and plants of many kinds. The book drastically changed thinking about pesticide and fertiliser use globally but it also spawned an awareness that the environment as a whole was under threat from largely unregulated farming practices. That in turn spilled over into the thinking of gardeners who wanted to 'do their bit' for the environment, and so wildlife gardening began. Nurseries sprang up selling wildflower plants and seeds and home gardeners started to grow them. It all came as a great surprise to many and I can recall a major national newspaper heading its gardening page with 'Weeds for sale'!

Then, church and churchyard authorities – or rather often, church support groups – drew parallels between the churchyards in their care and the wildlife gardening movement. Where they still existed as relics of nineteenth-century horticulture, formal areas of bedding plants were banished and in fact any resemblance to flower gardens was swept away. Churchyard grass, which for years had been closely and neatly cropped – labour was cheap and plentiful – was now at least partially left unmown. Today there are even a few churchyards designated as SSSIs [Sites of Special Scientific Interest] and these automatically gain some statutory protection, while rather more are designated SINCs [Sites of Importance for Nature Conservation] or SNCIs [Sites of Nature Conservation Interest].

Nonetheless, churchyard management for conservation is not the same as domestic gardening for wildlife. It is important to appreciate that wildlife gardening is about the creation of something that mimics a natural habitat but did not occur there naturally. By contrast, churchyard management for wildlife is principally about conservation – conserving what is there, rather than trying to create something that is not.

What, therefore, can churchyard authorities do to help preserve and conserve the unique nature of the land in their care? I shall offer here a basic summary of the issues that seem to me to be the most important. This is not and cannot be a comprehensive working handbook but it will, I hope, point those who are interested towards an appropriate way of thinking and give ideas they may wish to pursue in more detail elsewhere.

I cannot stress too strongly the importance of informing the parish of any intentions and seeking the comments and views of parishioners; strange as it may seem to those committed to conservation, some people prefer closely mown and neatly maintained churchyards. Then, and most importantly, investigate what information is available from old photographs and archived documents about the churchyard in times past. This should be followed by a careful survey of the plant and, if possible, the animal life present in the churchyard. There may be experienced naturalists living in the parish but it is likely that outside guidance will be needed; county naturalist trusts are almost always willing to help. I cannot emphasise enough the importance of always obtaining expert assistance – well meaning but misguided projects and enthusiastic amateurs can cause more harm than good and result in a wilderness rather than a wildlife reserve.

Given the information about the wildlife in your churchyard, a management plan should then be prepared; again expert help may be needed. However, it is important not to be impatient and imagine you can achieve all your objectives and realise all your ambitions within the first twelve months.

Most of the ground area in practically every churchyard is occupied by grass, sometimes of relatively recent origin, but sometimes an invaluable relic of ancient meadow. The bulk of this grass should be left unmown until late summer when the wild flowers will have set their seeds, leaving only mown pathways to facilitate grave visiting. Grass around the base of gravestones should be trimmed carefully by hand so the growth of shade and moisture-loving lichens on the stone is not adversely affected by exposure to full sun. By late summer and after mowing, the cuttings should be left for a week or so and turned occasionally – as hay is in farm fields. This will encourage the seeds to be shed before the residue is finally cleared away to the churchyard compost bins or offered to local horse owners who will often appreciate chemical-free hay.

If there is an old area of regularly mown grass that has never been treated with artificial fertilisers, treasure it and continue to mow it closely in the hope in due course of encouraging the beautiful and often rare toadstools known as waxcap fungi.

Use no pesticides or weed killers unless locally to treat a serious problem – spot treating invasive weeds such as docks or ground elder, for example – and apply no fertiliser to the grass, as this will encourage it to grow lush and lank, to the detriment of any wildflowers.

Resist the temptation to try and create a wildflower area by clearing a patch and sowing a wildflower seed mixture. Isolated patches of wildflowers in a churchyard, even assuming they will establish, can look anachronistic or plain silly. It may be possible to enhance the local churchyard flora with seed or plants of missing species but the introductions must be of known British provenance. A specialist wildflower nursery will give advice. However, do not become over-excited about the possibility of introducing wild orchids, as they are exceedingly difficult to establish in places where they do not occur naturally.

Where hedges do not exist along the churchyard boundary, plant new ones using a mixture of native species including beech, blackthorn, hawthorn, hazel and holly. Birds will feed on the fruits and nest in the hedge, while the hedge bottom will in time be colonised by herbaceous plants such as hedge mustard that favour this special habitat and will in turn attract their own characteristic wildlife.

Where the churchyard boundary is a wall, treasure it. If it is a dry-stone wall, treasure it still more because although they can be havens for snails, they are havens for a multitude of other small creatures as well. Many use them as temporary shelters for hibernation or nesting but for others they are permanent homes.

Countless species of creature will benefit – insects, spiders and other invertebrates, rodents like mice and voles, insectivores such as shrews, bats if the wall is big enough, amphibians (frogs, newts and toads will all use them) and reptiles (if you are especially lucky, you may have lizards in your churchyard wall). Many birds, too, will be attracted by dry-stone walls, partly as perches, partly because of the food they find on, in and nearby, but also as nesting sites – tits, robins, flycatchers and others will make nests in walls. The many crevices and ledges will collect small particles of wind-blown soil and before long mosses will colonise, to be followed by small ferns and flowering plants.

Where the churchyard boundary is merely a fence, do not despair, for fences offer a huge number of nooks, crannies and ledges; perfect places for insects and other small invertebrates to shelter, hibernate or breed. Examine almost any wooden fence more than about a year old and you will find a multitude of spiders' nests, pupae of butterflies and moths, and clusters of hibernating insects such as ladybirds. The greatest enemy of the wildlife that lives on and in your fence is wood preservative and paint. I can fully understand you will want to obtain the maximum life for your churchyard fence, so I suggest it is painted or treated once when it is first erected. Use a long-life non-toxic modern

wood treatment product and then, if at all possible, resist ever doing it again.

If the churchyard contains few or no trees, plant some, the number and kind depending on the space available but be sure not to end up creating a wood for a future generation to deal with! Although exotic trees have been planted in graveyards for centuries and any already present should remain and be cared for, I believe new plantings should be of native species. Oaks are good trees with which to start because more types of creature in Britain are dependant on, or make use of, our two native oak species than any other kind of tree. They are host to many hundreds of species of moth, beetle and other insect (more than forty are responsible for causing conspicuous and characteristically attractive galls), many kinds of spider and other invertebrates. Over 400 lichen species have been found growing on oaks, along with around sixty-five kinds of moss and liverwort and over 4,000 different fungi – albeit many microscopic – while of course many birds and several mammals are habitually associated with the trees.

Do not be too scrupulous in clearing away twigs and small fallen branches, as they are important habitats for many organisms, including numerous small animals, fungi and mosses. By similar token, minimise path treatments to those needed for public safety – such as clearing slippery algal growth – or for the removal of invasive weeds. Many

species of small flowering
plant, as well as some mosses and lichens,
grow harmlessly on and in gravel and brick paths.

Erect several types of bird nesting boxes in appropriate places in the hope of attracting a range of different species. Undertake some research into the various types of box available to discover how different bird species prefer boxes of varying overall size, shape, hole diameter, position and compass orientation. Organise hedge cutting to follow the time when nestlings will have flown, and do not disturb any existing natural bird nests – removing martins' or swallows' nests from buildings is sacrilege and according to tradition brings bad luck.

Planning and, even more importantly, listed building consent applications will invariably be assessed by the local planning authority for their conservation significance and churchyard authorities must be prepared to amend their plans to comply. Most commonly a bat survey will be required and this must be undertaken by an appropriate licence holder. If bats are found, they must not be disturbed or their roosting sites disrupted without official permission. Erect bat boxes to encourage summer roosting in particular; most areas have a local bat group of knowledgeable specialists who will give advice.

Repair of stonework should be done carefully, with minimal damage to lichen and mosses growing on it. Where re-pointing is necessary, some loss of species growing on the stone is unavoidable but experienced contractors will know this and work sympathetically. Do not clean headstones other than gently to render the inscriptions legible but do not scrape off lichens; remember these headstones may be their only habitats for miles around.

Finally, do not plant non-native flowers and shrubs in your churchyard – although they may attract pollinating insects while birds may feed on their seeds, my view is that they should be left to gardens where they are more appropriate and can be properly tended.

Forasmuch as it hath pleased Almighty God
of his great mercy to take unto himself
the soul of our dear brother here departed,
we therefore commit his body to the ground; earth
to earth, ashes to ashes, dust to dust;
in sure and certain hope of the Resurrection
to eternal life.

ENGLISH BOOK OF COMMON PRAYER, 1662

POSTSCRIPT

These are difficult times for churches and church authorities everywhere. There can be few areas where places of worship, especially it seems to me Christian places of worship and most especially perhaps those of the Church of England, are not facing crises of diminishing congregations and rising costs of maintenance. It is no part of my purpose to explain or try to suggest remedies for either – many knowledgeable authorities have already tried and failed – but the situation has repercussions for their churchyards.

Much has been achieved for rural Christian communities, their churches and churchyards over the past thirty years by the Living Churchyards and Cemeteries Project initiated by the organisation now called Germinate – The Arthur Rank Centre, an ecumenical Christian charity based at Stoneleigh in Warwickshire, not least in linking churchyard management and conservation with school education and the National Curriculum. The Centre has a far wider remit than churchyard conservation but has certainly been at the forefront of raising awareness. However, it should not be forgotten that not all churches and churchyards in need of help are either rural or Christian and there is in some ways an even greater need and scope for work in urban and inner city parishes of all denominations and faiths.

Of course, in an ideal world all places of worship would remain functioning for their originally intended purpose but if this is not financially practicable, possibilities exist for the transfer of the building to a conservation body. A number of excellent organisations undertake a protective role for churches, most importantly the Churches Conservation Trust and the strangely named Friends of Friendless Churches. (Not many churches I find are totally lacking friends but a vast number are painfully lacking funds!) These two bodies are constituted and funded in different ways and with slightly different remits but share a purpose in the protection and conservation of church buildings that are to a greater or lesser degree unused for worship, albeit many are still consecrated. The Churches Conservation Trust cares for over 350 English parish churches, while the smaller Friends of Friendless Churches owns around twenty-five churches in England and twenty-five in Wales. Smaller still, the Scottish Redundant Churches Trust cares for seven in Scotland, while the Ulster Architectural Heritage Society does not own but offers advice to the owners of historic buildings – including churches – at risk, and in the Republic of Ireland some state funded grant aid is available for those churches that are classed as 'protected structures'. Inevitably and in all cases, however, funds and resources seldom stretch to much more than making the buildings weatherproof. Most of the protected churches have an associated churchyard but that generally tends to fall – literally and financially – outside the scope of any help.

However, every cloud has a silver lining, albeit a slightly elusive one. Even if churches must close, their burial grounds need not be neglected. Disused churches do not have to be surrounded by an unkempt wilderness and I am reminded again of the statistic I cited in an earlier chapter: that around 6,000 churchyards in England are today managed specifically for their wildlife interest.

I recently and quite by chance found a remote and beautiful but largely redundant church in rural Wales that perfectly exemplified the situation. The church was in the care of the Friends of Friendless Churches. It had been re-roofed and was still used occasionally for services during the summer months. There was an old and evocatively picturesque graveyard running up a steep slope alongside the church building and an elderly parishioner was strimming the grass. He told me he had been born and lived his life in the nearby village and undertook the work occasionally, mainly because his wife and other members of his family were interred there and he wanted the churchyard to 'look respectable'. His motive and activities were both understandable and praiseworthy but it seemed to me that with some directed management and an input of expertise – and of course a willing body of volunteer labour

– the site could become more valuable and meaningful as a wildlife sanctuary. The man told me there were already bats roosting in the church and I have no doubt that a flora and fauna survey would reveal there was already an important biodiversity in the graveyard; I spotted a significant number of wild flower species, some fairly uncommon. Access to the graves could still be achieved by mown paths and the graves themselves could be maintained as dignified memorials without completely sanitising the headstones. However, all this unarguably worthy objective would be dependant on individuals giving up their time and perhaps a small amount of money.

Guidance for such projects abounds in the literature and on the websites produced by a wide variety of charitable and other organisations at parish, diocesan and national church level, while many county wildlife trusts and specialist scientific societies have produced their own basic information. If this book manages to inspire just one more churchyard group to take advantage of the opportunities offered by their local church, whether it is disused or still fully functional, and if just one more church authority lends its support, I shall personally be delighted. However, I hope and rather suspect there may be many more.

FURTHER READING

Child, M. *Discovering Churchyards*
   (London, Shire Publications, 1982)

Cocke, T. [ed.] *The Churchyards Handbook* 4th edn.
   (London, Church House Publishing, 2001)

Cooper, N. *Wildlife in Church and Churchyard* 2nd edn.
   (London, Church House Publishing, 2001)

Cornish, V. *The Churchyard Yew and Immortality*
   (London, Frederick Muller, 1946).

Dobson, F. *Guide to Common Churchyard Lichens*
   (Shrewsbury, Field Studies Council, 2004)

Gregory, D. *Country Churchyards in Wales*
   (Llarwst, Gwasg Carreg Gwalch, 1991)

Hackman, H. *Wate's Book of London Churchyards*
   (London, Collins, 1981)

Lees, H. *Cornwall's Churchyard Heritage*
   (Chacewater, Twelveheads Press, 1996)

Lees, H. *Hallowed Ground: Churchyards of Gloucestershire and the Cotswolds* (Parkend, Thornhill Press, 1993)

Lees, H. *Hallowed Ground: Churchyards of Wiltshire* (Bridgend, Picton, 1996)

Love, D. *Scottish Kirkyards* (Darvel, Alloway Publishing, 1989)

Redgrave, S. *Here Lyeth the Body: A Look at Worcestershire Churchyards* (Bromsgrove, Halfshire Books, 1992)

ACKNOWLEDGMENTS

I am extremely grateful to Lord Harries for his most thoughtful foreword and to my long-time friend the Rev John Samways for his constant interest and encouragement.

At Unicorn Publishing I was delighted to find Simon Perks was immediately receptive to my idea and saw the potential of the subject while Felicity Price-Smith then used her design skills and self-evident artistic talent to wonderful effect and turned my text into a work of enduring beauty.

Published in 2018 by Unicorn
an imprint of Unicorn Publishing Group LLP
101 Wardour Street
London
W1F 0UG
www.unicornpublishing.org

Photography Credits
p8 & p79 © Stefan Buczacki
p20 © Craigy144 at English Wikipedia
p32 © Rob Bendall
p73 © Acabashi, Creative Commons CC-BY-SA 4.0, Wikimedia Commons
p118 based on a photograph by The Gentle Author
All other photographs © Felicity Price-Smith

ISBN 978-1-910787-74-8

10 9 8 7 6 5 4 3 2 1

Designed by Felicity Price-Smith
Printed in India by Imprint Press